Duty and Conflict
Operation Iraqi Freedom

Vickey Joins the Military Series

Book 8

Duty and Conflict
Operation Iraqi Freedom

Book Eight

Table of Contents

Routine Combat Rotation

Journey to Iraq.

I had been at my new duty station in Germany for a few months after my return from Saudi Arabia, but for me and many others the war on terror, Operation Iraqi Freedom (OIF), began during that routine rotation in Saudi. It was the act of terrorism on American soil—the September 11, 2001, attack on the World Trade Center in New York City—that ignited the conflict for us.

The Patriot force of the Army had arrived in Southwest Asia over twelve years prior and remained ever since; their initial arrival coincided with Operation Desert Storm. Consequently, my unit was just one among many that had undergone a routine rotation in Riyadh, Saudi Arabia. But our rotation turned out to be anything but routine.

In the region, tense moments were always present, yet the events of September 11, 2001 (9/11), amplified more than just tensions; they ignited immense anger. The actions of that fateful day on America spurred the call to mobilize Patriot forces once again, to establish a forward presence in response to escalating tensions.

Following the attack on 9/11, myriad threats and accusations abounded, though none materialized—at least, not until over a year later.

Prior to the buildup of forces in support of Operation Iraqi Freedom within the Central Command area of responsibility, my unit and numerous other forces were already deployed in the theater, fulfilling rotations for Operation Southern Watch. Patriot troops were stationed at various air bases across the region, while others carried out routine missions in Kuwait, safeguarding national assets in Southwest Asia.

However, due to the increasing likelihood of an impending offensive operation in Iraq, our Patriot command opted to initiate a buildup of forces in the region. It wasn't until September 2002, though, that the US commenced the construction of combat power, simultaneously deploying additional forces to Kuwait in anticipation of potential conflict with Iraq if demands remained unmet. The initial deployment orders for an air defense unit, beyond those of us already present in the region for the standard Southwest Asia rotation, were issued at the end of September.

Eventually, Operation Southern Watch underwent a transformation; what was once a routine surveillance mission evolved into an overwatch mission following the events of September 11 in Afghanistan. Consequently, it was rebranded as Operation Enduring Freedom. US military forces and Patriot elements diligently prepared for conflict—all that awaited were the orders to proceed, none of which had materialized yet.

The deliberate buildup of US forces persisted until January 2003. Suspicions loomed over Saddam Hussein's alleged connections with Osama Bin Laden, the suspected mastermind behind the entire terrorism attack. President George W. Bush sternly demanded that Saddam cease his production of weapons of mass destruction. Despite Saddam's claims of possessing no such weapons and his refusal to comply with United Nations regulations, he was issued several ultimatums. However, a final ultimatum was extended, stipulating compliance by March 20, 2003, failing to comply would ensure that military action would ensue. With Saddam's non-compliance, America once again placed its trust and confidence in its military.

Deployment Orders

February 2003.

After countless months of waiting, working, and preparing our unit for possible deployment to Kuwait and subsequently to Iraq in support of operations in Southwest Asia, my unit finally received deployment orders; for us, that was when the real work began.

The deployment orders necessitated relocating our base of operations from Germany to the conflict zone, a task involving numerous logistical challenges. We were directed to utilize rail and sea transportation resources for the move. Our next objectives revolved around transporting and loading our unit's weaponry arsenal—a task that many of us had never before undertaken.

Commencing 24-hour operations, we diligently loaded hundreds of pieces of equipment without a single safety incident. Within a week, the entirety of our unit's equipment was successfully loaded onto rails and ships, bound for the Aerial Port of Debarkation in Kuwait. As we wrapped up that monumental operation, a collective sigh of relief echoed through the ranks; we had achieved a feat that many of us had never before attempted, leaving us all with a profound sense of pride.

Less than a month later, those of us slated for deployment found ourselves en route to Kuwait. In a swift turn of events, elements of the Battalion began departing from Germany to the combat operations zone on March 5, less than a month after receiving deployment orders.

After all the anticipation and preparation, the time had finally arrived for me and my unit to head to the battle region. March 17, 2003, held a significance unlike any other day for the personnel of my unit in Germany. On that day, we would depart from our home base en route to Kuwait. It was a momentous occasion, one that we knew would be etched in our memories just as previous generations heading to conflict zones had remembered their deployment dates.

The night before our departure, sleep eluded me; I tossed and turned, eagerly awaiting the alarm that would signal the beginning of what promised to be a memorable day. Unable to rest, I rose from bed and gazed out the window. The tranquility of the night was occasionally interrupted by the gentle tapping of tree branches against the window, as if seeking entry. The soft breeze caused the branches to sway rhythmically, captivating my attention. Perched on the second floor, I felt no concern about leaving my curtains open at night, a luxury I wouldn't have afforded myself back in the States.

With a slight veil of dew adorning the windowpane, I continued to gaze outward, embracing the tranquility of the moment and absorbing the silence that enveloped me.

Suddenly, “BUZZZZ.”

Startled, I swiftly silenced the blaring alarm. The bright, blue fluorescent glow of the alarm clock numbers cast an otherworldly hue across the room. The time—12:15—seemed to radiate with an intensity sharper than usual. However, there was no time to linger; I had to report to the unit by 2:00 am, where personnel accountability recall, and baggage line-up were scheduled for 4:00 am.

By 6:00 am, the buses had arrived to transport us to the terminal, where we would catch our flight—a six-hour journey to Kuwait, a layover before our onward journey to the war-torn country of Iraq.

On the plane, I settled into my seat, allowing myself to sink into its comfort. I rested my head against the headrest and closed my eyes, seeking a moment of tranquility amidst the impending journey.

Silence with a background "Hmmmmm" filled the air. "Hmmmmm." The silent sound persisted for several minutes, gradually giving way to the escalating roar of the engines as the plane commenced its slow, methodical taxi down the runway in preparation for departure.

Once the flight settled into its smooth ascent and reached its cruising altitude, I observed the flight attendants bustling about. However, exhaustion weighed heavily upon me after a long day of preparations, so I surrendered to the gentle lull of the aircraft and allowed myself to drift off to sleep.

Awakening a few hours later, I indulged in a bite to eat, followed by an in-flight movie and several games of solitaire. Before I knew it, the announcement rang out, signaling the impending descent and preparation for landing.

The plane touched down smoothly, gliding onto the tarmac before coming to a gentle stop. As we gathered our belongings and began to disembark, the weight of our personal equipment felt palpable. Stepping off the aircraft, I was greeted by a view unlike anything I had ever seen before. Stretching before me was a vast expanse of sweltering sand, with waves of heat shimmering in the distance. We had arrived in the deserts of Kuwait.

Arrival in Kuwait

The rallying station.

Upon our arrival in Kuwait on a Tuesday morning, we were swiftly transported to Camp Begin, a designated rallying point for incoming personnel to the region. As we reached the camp, we received a concise briefing before being directed to a holding tent. There, we were instructed to await the arrival of the rest of our personnel, the delivery of our equipment to the region, and the issuance of further orders—orders that would dictate our potential entry into Iraq, should the need arise.

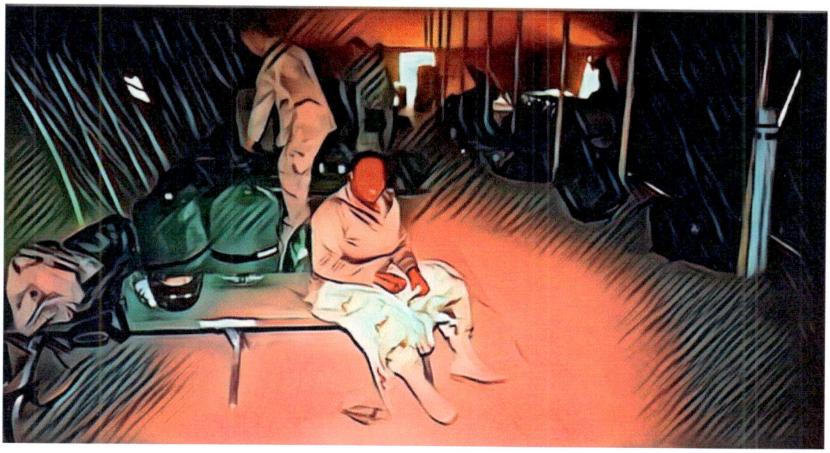

Within days of our arrival, the ship carrying our Patriot weapon system and supporting equipment docked at port. Subsequently, unit personnel made their way down to the port to commence convoying vehicles back to Camp Begin. Despite our continued state of readiness and anticipation, we remained in a holding pattern, awaiting further instructions. At that juncture, there was no concrete indication that a battle was imminent; nonetheless, we persisted in our preparations as if one were on the horizon.

However, on the day designated for transporting our equipment to camp, we were met with a formidable obstacle—a massive sandstorm. That natural phenomenon thwarted our efforts to ferry all the equipment from the port to Camp Begin in a single day. While many of us had previous experience with dust storms from our time stationed in the desert lands of El Paso, Texas, nothing could have prepared us for the intensity of the sandstorm we encountered on that day in Kuwait.

We later learned that sandstorms in the region were a common occurrence and could pose significant challenges during battle, a reality we experienced firsthand long before setting foot on the battlefield. The blowing sand, which drastically reduced visibility, brought our operation to a standstill for an entire day.

As we traversed the desert en route to Camp Begin, transporting our weapon system, we encountered further obstacles. The soft sand, coupled with the weight of our equipment, led to many vehicles becoming mired in the terrain. The sheer mass of our ton-sized equipment caused it to sink deeply into the fine particles of sand. It became apparent that our equipment was ill-prepared to contend with both the sandstorm and the extreme temperatures that fluctuated throughout the day. Mornings and evenings brought chilly temperatures, particularly in the early days of our arrival in the country, necessitating the use of cold-weather gear. Conversely, the afternoons brought sweltering heat, testing the endurance of personnel.

Within days of completing the pickup of our equipment from the port, we were issued live M16 ammunition—an unmistakable sign that we were indeed on the trajectory towards battle. Initially, the receipt of live ammunition was a bit daunting for me, as I had never handled a weapon with live rounds. However, I took solace in the extensive training I had received over the years, trusting in my ability to handle the weapon safely despite the added element of live ammunition.

As days passed, deployment operations to the region continued. Each day, we found ourselves confined to our sweltering tent, eagerly awaiting orders that would dictate our next steps. Yet, despite the ongoing preparations, no orders arrived, and we remained no closer to movement towards Iraq.

As we waited for those elusive orders in our holding tent, with our equipment unloaded from the cargo ships and stationed in the Camp Begin area, we took proactive measures to ensure readiness for the battlefield. Testing the equipment and bringing it to operational status became our priority, preparing for the possibility of imminent deployment.

The Patriot equipment was meticulously set up and powered on during the testing phase. However, it was during that crucial evaluation that a sobering reality emerged—the systems were riddled with problems. A significant number of radar systems, vital components for detecting incoming missiles, were deemed non-mission capable. That revelation meant that critical assets would remain unprotected by Patriot forces unless the issues were promptly addressed and resolved.

But our teams didn't hesitate to spring into action, determined to rectify the issues plaguing our systems. The maintenance teams worked tirelessly, their efforts fueled by determination and a sense of duty. Meanwhile, our supply specialists exhibited unwavering resolve in sourcing the necessary parts for the repairs and ensuring those parts reached the right locations in a timely manner. Amidst the flurry of activity, there was also an intense sense of collective hope, accompanied by fervent prayers for success.

Through a combination of hard work, resourcefulness, and faith, the equipment was gradually restored to mission capable status. As the last repairs were completed, a wave of relief washed over us all. With our equipment operational and our assets now defended, we could finally breathe a collective sigh of relief, knowing that we were prepared to face whatever challenges lay ahead.

Days into our deployment and after ensuring our weapon system was operational, news reached us that the Central Command had initiated combat operations against Iraq. Despite being stationed in tent city at Camp Begin, we found ourselves thrust into the heart of the conflict, Operation Iraqi Freedom, on the morning it commenced.

Earlier in the week, President Bush had issued a 48-hour ultimatum to Iraq and Saddam Hussein to comply with specified conditions; failure to do so, they would face consequences. As Thursday, March 20, 2003, dawned, it became apparent that Saddam had once again failed to comply. In the early hours of March 20th, President Bush formally declared combat with Iraq. It was on that fateful day that Iraqi missile forces launched the first of more than twenty missiles at coalition forces.

When official combat operations commenced, some air defense forces were still arriving in the area of responsibility, and many units were awaiting the arrival of their equipment. However, in the days that followed, all personnel and equipment were swiftly deployed to the conflict area, and the next steps were imminent.

On the first day of combat, elements of Patriot units wasted no time, departing Camp Begin within hours of the operation's commencement. Those Patriot units already positioned in the region were poised to avert disaster when, at around 1:00 pm on March 20, an attack occurred in Kuwait. A tactical ballistic missile was detected heading towards coalition forces, and our sister Battalion from Fort Bliss, Texas, swiftly intercepted the missile with a direct hit. Shortly thereafter, at approximately 1:20 pm, another missile launch was detected, prompting a unit from another Battalion from Fort Bliss to intercept it. The Patriot defense system was ablaze, fiercely safeguarding our assets.

During each missile launch, we were required to transition to Mission-Oriented Protective Posture 4, mandating the donning of our complete nuclear, biological, and chemical gear, including masks. With temperatures exceeding ninety degrees Fahrenheit in Kuwait at the time, the added layers of protective gear exacerbated the already extreme heat conditions. Despite the discomfort, we remained vigilant as missiles continued to be launched throughout the day.

As nightfall descended, we found ourselves still clad in our chemical gear. The first twenty-four hours of combat operations proved to be some of the busiest shooting periods for air defense. Despite the challenging circumstances, our focus remained unwavering as we stood ready to defend against any further threats.

As the missile barrage persisted, we found solace in the advancements made in technology and tactical strategies over the past decade. Compared to our experiences during the Desert Storm conflict a decade prior, our Patriot units were now far better equipped to defend against missile threats.

Moments after the early warning sirens sounded, the distinct roar of Patriot missiles launching from their launchers echoed across the camp. That sound, familiar from countless live fire missions, drowned out the wail of the warning sirens. The renewed battle between Patriot and Iraqi missile forces had commenced, but that time around, we harbored a greater sense of confidence that Patriot would emerge as the unequivocal victor.

The morning of Friday, March 21, 2003, began with an eerie calm, but at approximately 11:20 am, we were jolted by the sound of booms overhead. Reacting swiftly, we donned our chemical gear in response to the threat. Later, we learned that the Kuwaiti Patriot unit had successfully engaged a missile near our location, just 20 to 25 miles from Camp Begin.

By Saturday, March 22, 2003, the atmosphere had settled, with only one missile alert throughout the day. Personally, I felt a profound sense of relief at the respite from the combat raging around us, as I had been battling illness since Thursday. Afflicted with what was colloquially termed the "Kuwaiti crud," I sought medication in order to ready my body for battle. It was imperative to me that I recover fully before any potential orders to move out were issued. Consequently, I made the decision to visit the clinic area, a step I would not have taken under normal circumstances, in the hopes of swiftly overcoming my ailment.

As the conflict continued to unfold around us, I remained confined to my cot, willing myself to recover. Amidst my illness, we received reports that approximately 8,000 troops from an Iraqi division had surrendered. However, the battle persisted, underscoring the ongoing intensity of the situation.

When Sunday arrived on March 23, 2003, we finally received the welcome news that we could remove our chemical gear. We had been encumbered by the stifling, hot protective gear for days, ever since the onset of combat operations on March 20th. Although we still had to carry the gear with us at all times, the reprieve from wearing it unless a warning siren sounded was a small comfort.

However, the moment of relief was swiftly overshadowed by devastating news— one of our sister units had mistakenly shot down a friendly aircraft, casting a shadow over the Patriot forces. The day took a turn for the worse as further distressing reports emerged. First, we learned that the commander of the unit responsible for the friendly-fire incident had been relieved of duty. Then, shockingly, news spread of a soldier within our ranks hurling a grenade in his tent, injuring some of his fellow team members. Finally, we received word that an element from another sister unit had become disoriented upon crossing into Iraq and subsequently came under attack.

Trapped within the confines of our tent area, we were privy only to fragmented updates, with no real contact with the outside world. Information was scarce, limited to what leadership chose to disclose to the troops. The lack of news, newspapers, or reliable intelligence left us feeling isolated and uncertain amidst the tumultuous events unfolding around us.

As March drew to a close, the conflict in Iraq showed no signs of abating. Despite the ongoing turmoil, we remained stationed at Camp Begin, steadfastly awaiting our orders to move into Iraq. Throughout that period, we maintained rigorous testing and preparations for our equipment, ensuring it was fully operational for the tasks ahead. Loading our launching stations with twenty live missiles added to the intensity of our readiness efforts.

Simultaneously, we made preparations for our impending departure from Kuwait, carefully selecting and loading personal equipment necessary for our mission in Iraq. Any surplus gear was stored in storage containers until our eventual return to the area. The directive to travel light underscored the urgency and gravity of the situation.

On Saturday, March 29, 2003, rumors swirled that we were scheduled to depart that evening, crossing into the live combat zone of Iraq. However, no official confirmation had been received. Amidst the uncertainty, we also heard murmurs that Operation Iraqi Freedom could potentially conclude within a week, a glimmer of hope amidst the prevailing uncertainty and anticipation.

Amidst the ongoing combat operations in Iraq, Patriot continued to demonstrate its effectiveness, instilling confidence in its capabilities. As Iraq persisted in violating the terms of the Desert Storm cease-fire agreement, the United States turned to Patriot to neutralize the threat. Remarkably, the weapon system consistently achieved direct hits with every release, effectively thwarting enemy attacks.

Throughout the relentless battle, Patriot emerged as a stalwart defender, thwarting missile volleys day after day. Its presence on the battlefield proved pivotal, disrupting Iraqi efforts and safeguarding coalition forces. The relentless barrage of enemy missile launches was met with swift and decisive action from Patriot air defense systems. Notably, one thwarted attack was credited with saving the coalition forces' command headquarters, underscoring the critical role played by Patriot in the ongoing conflict.

The long wait had finally come to an end. No longer relegated to the sidelines, we had received orders to advance from Kuwait into Iraq. Excitement coursed through our ranks at the prospect of joining the fray, yet beneath the surface, there lingered an intense sense of nervous energy and fear.

As the night unfolded, sleep eluded many of us, our rest disturbed by the distant rumble of intense firefighting. We were acutely aware that the coming days would bring heightened intensity and rapid movement. With our imminent entry into the battle zone, the reality of our role in the conflict loomed large, filling us with a mixture of anticipation and apprehension.

Crossing into Iraq

The movement forward.

Since our arrival in the country, our camp had been repeatedly rocked by the thunderous explosions of missiles detonating in the distance. The relentless barrage of missile attacks and counterattacks persisted for hours, extending into the evenings and stretching across multiple days. Throughout the ordeal, we were compelled to remain clad in our chemical gear, enduring the suffocating heat and discomfort with each passing attack cycle.

Day after day, we languished in our chemical gear, trapped in a holding pattern at the camp as we awaited the subsiding of the crisis. Yet, the interminable wait seemed to stretch on endlessly, exacerbating the already unbearable conditions. As the new month approached, temperatures began to rise, inching closer to the scorching one hundred-degree mark. The oppressive heat, amplified by the layers of protective gear, made each moment feel excruciatingly hotter. Even as night descended, there was little respite from the sweltering conditions, further compounding our discomfort and frustration.

As I lay on my cot, attempting to minimize any unnecessary movement, a sense of déjà vu washed over me. It was another instance of departing from the relative safety of our camp to venture into the uncertainty of the unknown. With the announcement that we would be leaving Camp Begin, Kuwait, the following day en route to the Iraq-Kuwait border, I reclined once more, closing my eyes in contemplation.

Suddenly, the tranquility was shattered by a deafening boom. "What was that?" someone exclaimed. It wasn't until later that we learned another barrage of missile attacks had occurred, serving as a stark reminder of the volatile reality surrounding us.

As I reflected on the soldiers from our sister unit who had been taken captive, I couldn't help but imagine the harrowing ordeal they must be enduring. However, my attempts to empathize were hindered by my lack of firsthand experience with captivity, my only frame of reference being what I had seen depicted on television. Realizing the futility of dwelling on their plight and the unknowns beyond Camp Begin, I made a conscious effort to quiet my thoughts and focus on the present moment. I suppressed the urge to entertain the "what ifs" that threatened to consume my mind.

As the hour of departure drew near, a sense of solemnity settled over us. It was time to bid farewell to the safety of Camp Begin. We embarked in convoy formation, battle gear at the ready, bracing ourselves for the journey ahead. The convoy journeyed for approximately three hours until we reached a designated stop, where we paused, preparing ourselves for the imminent crossing into the unknown territory of Iraq.

During the stop, just before crossing the border, we received a briefing outlining the plan for our next steps. Emotions ran high as we discussed our impending move, and it was clear that reality was beginning to sink in for many of us. Some individuals became visibly emotional, struggling to maintain composure in front of their fellow soldiers. Instructions were issued to compose themselves, especially in the presence of their comrades.

Despite anticipating that we would have ample time to mentally prepare for what lay ahead, the reality was quite different. We had no choice but to swiftly transition into preparation mode. In the ensuing hours, we fueled up the vehicles and grabbed a quick bite to eat, all while knowing that the clock was ticking.

As night fell, we loaded up our gear and embarked on our journey toward the Iraqi border. Within a matter of hours, we had crossed from Kuwait into Iraq, officially entering the conflict zone. With that pivotal move, we took our first steps into the heart of the battle, officially becoming integral participants in Operation Iraqi Freedom.

Movement Within Iraq

Fighting sites.

Following our initial crossing into Iraq, we awaited the arrival of an escort element before proceeding further. From that point onward, our unit would be accompanied by armed escorts for any movement within the region. With a military escort at our side, flanked by M1 battle tanks and Bradley fighting vehicles, we set out toward our initial objective—the site of our first engagement.

As we traversed the desert terrain, the distant echoes of mortar fire reverberated through the air, a constant reminder of the volatile landscape surrounding us. However, the exact proximity of the source remained elusive, shrouded by the vast expanse of the desert sands.

As we traversed the route, we encountered remnants of forces that had preceded us, indicating some level of clearance along the path. However, the journey proved exceedingly arduous, especially for our equipment. The terrain was unforgiving—rugged, unpaved, and recently cleared for travel. Despite the challenges, reaching our first fighting site was paramount, as our progression would enable other units to advance to strategic locations, bolstering the defense of multiple US assets.

The responsibility of our Patriot higher echelon was immense. They had to strategically deploy enough Patriot forces to safeguard assets in Kuwait, Bahrain, Qatar, Jordan, and Saudi Arabia, all while ensuring sufficient protection for our maneuver forces operating within Iraq. It was a monumental task, and our unit played a crucial role within that complex operation.

Undeterred by the harsh conditions, our determination to reach our objective remained unwavering. The maneuver of Patriot units to the frontline marked a historic shift. Prior to the Iraq conflict, Patriot had always functioned as a rear asset, but as we pressed forward through the rugged terrain towards the frontline, we were charting new territory in the annals of military history. The Iraq battle would solidify Patriot firing units as frontline defenders—a significant milestone in our unit's legacy.

The journey unfolded like a scene from a movie, with thick, billowing clouds of dust trailing behind each large vehicle, obscuring the vision of those forced to travel behind. The procession of over one hundred vehicles traversing unpaved, gravel-strewn terrain made our unit a conspicuous target for the enemy. The deafening roar of engines, characteristic of a convoy of 5- to 10-ton vehicles, echoed across the landscape, impossible to conceal.

As our convoy cautiously advanced along the partially cleared route, tension hung thick in the air. With weapons at the ready and vigilant eyes scanning the surroundings, we remained on high alert for any signs of enemy activity. The unmistakable trail of dust and the thunderous rumble of our vehicles made our presence difficult to conceal, painting a clear target for potential adversaries. As we forged ahead, heading directly into the line of fire, our convoy became a moving target, vulnerable to enemy engagement.

Upon reaching our first resting point safely, the reality of our situation in Iraq began to sink in, heightening our sense of apprehension. Unit commanders swiftly devised guard shifts, ensuring a constant vigilance over our surroundings. As we disembarked to stretch our legs, tend to our equipment, and take much-needed rest, we were keenly aware of the ever-present threat lurking beyond the confines of our temporary refuge. Patrols were organized to secure the area, a necessary precaution in the volatile environment of a combat zone.

During our stop, the realization dawned on many of us that opportunities for routine bathroom breaks would be few and far between, if they existed at all. We had to devise a creative solution to address the pressing need as our journey continued. Nothing in our training had prepared us for that real-world challenge, but adaptability was key as we brainstormed ways to manage the essential aspect of daily life on the move.

We recognized that the nature of our travel meant that conventional methods of finding restroom facilities would not apply. As we contemplated solutions for future moments, one innovative idea emerged: modifying the seat cushions to create makeshift waste disposal systems. By cutting out portions of the cushions and inserting containers along with rubber tubing, we devised a practical solution that allowed waste to be safely expelled from the vehicle—an ingenious adaptation to our unorthodox travel circumstances.

Navigating the complexities of combat travel was a learning experience for us all, and we were continuously refining our approach as we encountered new challenges along the way.

Throughout the night, a palpable sense of fear and anxiety hung heavy in the air, casting a somber mood over our surroundings.

As I embarked on patrol duty under the cover of darkness, the eerie silence amplified my senses, heightening my awareness of the distant echoes of combat. In the distance, the flickering flares of battle illuminated the night sky, their brilliance reminiscent of a vibrant fireworks display on the Fourth of July. However, there was no jubilation or festivity to be found in those luminous bursts; rather, they served as stark reminders of the relentless cycle of death and destruction unfolding on the battlefield.

As my patrol shift drew to a close, a serene calm began to envelop the night, gradually dispelling the shadows of fear and uncertainty. With each passing moment, the tranquil stillness of the night gave way to the gentle warmth of the approaching dawn, infusing the air with a sense of peace and hope for the new day ahead.

"How can something so beautiful exist amidst this realm of destruction?" I mused, captivated by the ethereal beauty of the desert landscape unfolding before me. As the first light of dawn cast its gentle glow upon the glistening sand, a serene tranquility settled over the desert, momentarily masking the harsh realities of our surroundings. Despite the coolness of the morning air, a sense of calm pervaded the open expanse, offering a brief respite from the chaos of conflict.

The sudden reverberation of mortar rounds in the distance shattered the tranquility, jolting me back to the harsh reality of our situation. That it was not a tranquil paradise; it was the unforgiving terrain of Iraq, and my unit was navigating through the heart of a conflict zone. With a renewed sense of purpose, I completed my patrol and returned to my vehicle to prepare for the journey ahead. The convoy awaited our departure, poised to fuel up and advance to our first engagement site mere miles away.

The unfolding events seemed straight out of a screenplay, the kind of dramatic moments that inspire movie scenes. Yet, as mesmerizing as it was, the reality of our situation was daunting. Patriot units had never before maneuvered alongside a Marine Expeditionary Force or provided direct support to Marine forces, presenting a unique set of challenges as we embarked on that unprecedented journey together.

Navigating through uncharted territory, both literally and figuratively, we faced uncertainties at every turn. Would our equipment withstand the rigors of frontline combat? Could we keep pace with the Marines as we ventured into the heart of the conflict zone? Those questions loomed large, but amidst the chaos, there was no time for hesitation or doubt. With our departure imminent, we had to press forward, ready to confront whatever challenges lay ahead.

Until that moment, our journey had been largely characterized by the vast expanse of desert landscape stretching endlessly before us. However, amidst the barren terrain, we encountered a poignant reminder of the human toll of war: displaced Iraqi citizens, wandering aimlessly along the roadside, their uncertain gaze reflecting the upheaval of their lives.

Remarkably, our passage had thus far been devoid of hostile encounters or threats to our safety. As we embarked on the new day's journey at 8:00 am, optimism filled the air, buoyed by the hope of reaching our destination unscathed. However, our progress was soon hindered by unforeseen challenges.

A prolonged stop at a desolate outpost, aptly dubbed a "hellhole," consumed several hours of our time, delaying our departure until 3:15 pm. Determined to forge ahead, we resumed our journey, only to be met with a series of vehicle malfunctions that tested the resilience of our equipment in the face of adversity.

The breakdown of our parts vehicle, the Small Repair Parts Trailer (SRPT), dealt a significant blow to our logistics capabilities. Initially, we entertained the possibility of towing the trailer along with us, but it soon became apparent that that was not a viable solution. Reluctantly, we faced the reality that we would have to abandon the trailer, but not before salvaging its precious cargo.

With hundreds of critical parts essential for the maintenance of our Patriot system and vehicles housed within the trailer, there was no room for hesitation. We swiftly embarked on the arduous task of unloading the contents, ensuring that each component was carefully accounted for and transferred to our convoy. Every bolt, every circuit, every essential piece was meticulously gathered, a testament to the importance of maintaining our weapon system amidst the challenges of the battlefield.

As we resumed our journey, determined to overcome the setbacks that had befallen us, we soon encountered further challenges plaguing our convoy. The Battery Maintenance Center (BMC), vital for storing parts and system manuals, required additional hauling, adding to the strain on our already burdened resources. Shortly thereafter, the Electric Power Plant (EPP), responsible for powering the Patriot System, experienced gas pedal pressure issues, further impeding our progress.

To compound our woes, the 10-Ton Heavy Expanded Mobil Tractor Truck (HEMTT) – the workhorse tasked with pulling the launching station – succumbed to mechanical failure, halting our advance like a falling domino in a chain reaction. Despite our best efforts, the relentless onslaught of obstacles forced us to make difficult decisions. Ultimately, we were compelled to leave behind two pieces of equipment – the incapacitated HEMTT tractor and our indispensable parts trailer – as we pressed onward towards our first battle site.

As night descended upon us, our journey was further plagued by adversity. A sister unit encountered grave misfortune, losing a small trailer and a launcher station carrying four live missiles. Throughout the day, the unit had faced numerous other complications, leading to multiple halts along the treacherous roads of Iraq, notorious for ambushes and hostile activity.

Compounding our predicament, our escort was compelled to depart without us, unable to delay their onward mission any longer. It became painfully evident that the arduous task of transporting the

Patriot weapon system across vast expanses of unforgiving terrain, particularly the soft, treacherous sands of the desert, had presented unforeseen challenges. The sheer weight and bulk of the Patriot equipment proved no match for the relentless assault of the desert landscape.

Exhausted from the incessant mechanical problems and ceaseless vigilance, we reluctantly sought respite in a hazardous environment, risking our safety to steal a few precious hours of rest amidst the chaos of war.

Arriving at our designated battle site, we swiftly assumed our duties, taking over from the unit previously stationed there. With practiced efficiency, we transitioned our equipment into battle-ready configuration, meticulously testing each component to ensure its operational readiness.

However, amidst the urgency of our mission, we couldn't ignore the stark reality of our surroundings. The site bore the scars of previous conflict, a grim reminder of the tumultuous events that had transpired before our arrival.

As we settled into our new post, surrounded by the vast expanse of desert, we found ourselves stripped of the comforts and amenities we had grown accustomed to during training exercises. In the harsh reality of battle, such luxuries were deemed inconsequential, relegated to the sidelines as we focused on our primary objective. Though the site had been cleared to the best of their ability by those who had come before us, the harsh conditions served as a stark reminder of the challenges that lay ahead.

Adapting to the harsh conditions of the battlefield, we reverted to our fundamental soldiering skills, relying on ingenuity and resourcefulness to navigate the challenges we faced. Despite my years of service, that was the first time since basic training that I had to utilize my entrenching tool, a testament to the unique demands of the situation.

With no infrastructure in place, we had to rely on makeshift solutions for basic necessities. Digging cat holes for waste disposal and using baby wipes for hygiene became routine practices in our daily lives. The absence of shower and latrine facilities meant that we had to make do with the limited supplies we carried with us, adapting to the harsh realities of the open desert.

As the battle unfolded, the logistical trails were still in the process of being established, presenting a significant challenge for the engineers tasked with supplying and supporting our units. The vast distances we covered required extensive logistical support, with engineers working tirelessly to reach the various fighting sites and construct basic facilities. Keeping pace with the movements of smaller unit elements added another layer of complexity to the logistical mission, underscoring the immense challenges faced by our support personnel in sustaining our operations in the field.

The harsh reality of the desert environment came crashing down on us with the arrival of the sandstorm, eclipsing our concerns about hygiene. The ferocious winds tore through our site with unrelenting force, forcing us to seek refuge within the confines of our weapon system vehicles. The intensity of the storm rendered visibility virtually nonexistent, effectively nullifying our ability to conduct air defense operations.

In the midst of the storm, we were reminded of the unforgiving nature of the desert, where survival often hinged on our ability to weather the elements. Despite our best efforts to maintain operational readiness, nature had other plans, reminding us of the formidable challenges posed by the environment in which we operated.

The combination of the sandstorm and rain escalated the hazardous conditions at our battle site. Movement within the area became challenging, hindering our ability to navigate the terrain and coordinate with supply sources. Beyond the immediate dangers posed by the storm itself, the humid and hazy atmosphere heightened the risk of attack by Iraqi forces.

In response to the deteriorating conditions, units were directed to cease movement and fortify their positions. Hunkering down became the order of the day as we waited for the storm to abate and visibility to improve. Amidst the chaos of the storm, maintaining vigilance against potential threats remained paramount, underscoring the ever-present danger of operating in a hostile environment.

Later, the directive to move forward to Baghdad signaled a significant shift in our mission. Leaving our initial fighting site behind, we prepared to embark on the next phase of our journey into the heart of Iraq. As we readied ourselves for the move, the gravity of the situation weighed heavily on our minds.

Transitioning from one battlefield to another, we remained vigilant, maintaining a watchful eye on the skies for potential threats while fortifying our position against potential attacks. With each step forward, we inched closer to our ultimate destination, fully aware of the challenges and dangers that lay ahead on the road to Baghdad.

Next Moves Within Iraq

Road to Baghdad International Airport.

The shift in protection objectives to the Euphrates River marked a strategic adjustment in our mission. As Patriot units continued to make history with their forward engagements, our focus pivoted towards safeguarding vital assets along the Euphrates River.

With the new directive, our unit prepared to move once again, that time towards the designated protection zone. As we geared up for the journey ahead, we understood the significance of our role in defending critical infrastructure and maintaining air defense capabilities in the evolving battlefield landscape.

Despite the setbacks, our determination to press forward remained unwavering. With our convoy once again on the move, we navigated through the darkness aided by our night-vision goggles, a technology that proved invaluable in such challenging conditions.

As we encountered yet another round of mechanical issues, our resourcefulness came to the forefront. Through innovative solutions and teamwork, we managed to address the HEMTT starter problem and resumed our journey. Despite the delays, our resolve to reach our destination remained resolute, driving us forward despite the obstacles in our path.

The news of engineer support, while certainly welcomed, came at a time when we were already physically and mentally exhausted. The continuous combat operations and the transition into securing an entire division area added to our fatigue. Unlike other situations where units could be rotated to provide rest for personnel and equipment, the nature of the Operation Iraqi Freedom conflict presented unique challenges. With our Patriot units dispersed over long distances and unable to easily replace one another, the strain on both personnel and equipment became increasingly apparent.

Despite the exhaustion, we remained dedicated to our mission, knowing that our efforts were crucial in securing the area and supporting the overall objectives of the operation. We pressed on, drawing strength from our camaraderie and the knowledge that our contributions were making a difference on the battlefield.

The anticipation of reaching our next objective, the Baghdad International Airport, helped us push through our fatigue as we continued forward. That move was accompanied by armored escort, provided by the engineer battalion. The presence of those escorts was vital, given the expectation of encountering enemy contact during our journey. Whether in the form of direct or indirect fire, the support of the armored escorts ensured the safety and security of our convoy as we made our way towards our destination.

We would linger at the second site for a while before proceeding to the airport. The need to relocate to the airport stemmed from a perceived threat from another nation. It was believed that the country possessed some of Iraq's concealed weapons of mass destruction, prompting our departure to the airport.

Our day started early, at 4:00 am, with the packing up of the site. By 7:30 am, all equipment was organized and prepared for departure. Although the anticipation of what was to come was palpable, it gradually diminished as we embarked on the next phase of our journey.

The next phase of our journey was again disrupted by mechanical issues. Right after leaving the site, a Fire Control

HUMVEE broke down and needed towing. Roughly fifteen miles further, a severe collision occurred involving key components of our combat system: the Antenna Mast Group (AMG), the Engagement Control Station (ECS), and the Electric Power Plant (EPP). The accident necessitated the towing of both the ECS and AMG.

Subsequently, another unit in our convoy moved ahead, leaving us to deal with additional vehicle breakdowns. By 4:00 pm, when we were supposed to reach our destination, we were still not there; miles remained on our journey. It was indeed a challenging day. Compounding our troubles, the Battalion Commander and Battalion Sergeant Major were part of our convoy and witnessed the entire ordeal. Although it was an unfortunate sight, we were relieved that there were no injuries.

When our journey through Iraq initially commenced, we were accompanied by a military police escort, and the Infantry Division had already started to clear additional routes. However, in the final miles toward Baghdad, we were left to navigate on our own due to our escort's inability to wait while we addressed our mechanical issues. Although units like the 82nd Airborne, the 101st Airborne, the 4th Infantry Division, and the 3rd Infantry Division had cleared paths into

Baghdad, we had to depend on our navigation skills to reach our next battle site. Admittedly, my own navigation abilities were notably weak, leaving me to hope fervently that my Commander's skills were significantly stronger.

The accident further compromised our initial mission in Baghdad. Our unit, having worked tirelessly to keep our system operational and mission-ready, was expected to be the foremost unit upon reaching the airport area. Unfortunately, following the incident, our sister unit took over the role as the Battalion Commander's confidence in our Battery Commander to successfully execute the mission had diminished.

Upon reaching the Baghdad Airport area, we finally arrived at our combat site, but it was a disheartening day for our unit. We had diligently maintained our weapon system to ensure its operational readiness for the airport mission throughout our deployment, yet the vehicle accident resulted in our assignment to a secondary location. Despite the disappointment of being placed at a less critical site, we had to quickly regroup and establish our combat readiness.

After configuring our weapon system for combat, it would take another day for the engineers to arrive and enhance our site's defenses. They were scheduled to construct perimeter security mounds, a wooden latrine, concertina wire barriers, and a makeshift shower. Fortunately, the supply system had significantly improved since the battle's early days, enabling the engineers to more effectively reach various sites and fortify them. Meanwhile, our duties included guarding our site and managing a burn pit to dispose of our own garbage and waste.

Later, we learned that the site nearest to the Baghdad Airport, our original assignment, was no better than the one we were tasked with defending. The news brought us some hearty relief, helping to alleviate the disappointment of missing out on the higher-profile position.

We continued our air defense mission, safeguarding the skies for many additional days. Eventually, an order was issued to lower the air defense warning status from high to medium within Iraq for missile threats. That adjustment was significant, signaling a substantial decrease in the threat level from previous weeks. Despite the reduction, it also affirmed that our Patriot units would persist in providing protection for critical assets within the theater of responsibility.

Cease Fire

Post-Hostilities.

We reached a pivotal moment in the hostilities where our units could finally adopt a nuclear, biological posture zero. That meant we no longer had to wear or carry our chemical equipment constantly, although it needed to remain within ten minutes' reach. The update was particularly welcomed not just because the gear was cumbersome, whether worn or carried, but also because it indicated a significant reduction in the nuclear threat. That allowed us to breathe a collective sigh of relief.

Additionally, the air defense warning level was lowered from medium to low, marking more positive developments. Both the reduction in chemical posture and the decrease in air defense warning status served as clear indicators that combat hostilities were beginning to wind down, at least for our Patriot units.

By the end of April 2003, several units received orders to transition to alert state zero, which allowed them to power down their weapon systems and prepare for orders to return first to Kuwait, and then to their home duty stations. However, our situation was different. Due to the timing around Saddam Hussein's birthday, a period marked by heightened tensions, our equipment remained active and on high alert.

On May 1, 2003, the President officially declared an end to combat operations, but the declaration did not immediately translate to a cessation of danger. Despite the announcement, we remained stationed in Baghdad, an area where risks and uncertainties persisted.

The journey of Operation Iraqi Freedom led us from Camp Begin in Kuwait to the Baghdad International Airport in Iraq. We were actively engaged in the conflict zone from the first day the US initiated combat operations until the day the end of those operations was declared. With the primary objectives of the Patriot mission seemingly fulfilled, the focus began to shift.

As tensions in the region decreased, by the 21st of the month, our unit had transitioned into training mode. Our crews commenced training for their air defense certifications, taking the time to refine and hone their skills. The shift not only helped in maintaining operational readiness but also ensured that personnel stayed proficient and prepared for any future requirements.

By the end of the month, our unit was eagerly awaiting orders to depart Iraq and escape the relentless scorching heat. The combat operations of Operation Iraqi Freedom were winding down for the Patriot units, marking the close of a successful joint and coalition operation characterized by speed and agility. Despite encountering some obstacles along the way, our unit executed our tasks with precision; our mission achievements were underscored by displays of courage and resiliency.

The seamless integration of US and Kuwaiti Patriot firepower during the conflict showcased the effectiveness of coalition operations, proving that all the prior joint training had indeed paid off. The successful conclusion of Patriot operations in Operation Iraqi Freedom confirmed that the twelve years of US Patriot rotations in and out of the region following Desert Storm had been instrumental in our success. Additionally, air defense soldiers had the opportunity to refine their expertise in joint combat operations, further enhancing their capabilities.

Historically, the role of air defenders had been crucial in ensuring the protection of military forces. Valued for their multifunctional capabilities, air defenders had consistently provided essential protection against both ground and air attacks. During the advance to Baghdad, air defenders demonstrated their capacity to act as a force multiplier; they not only advanced forward but also engaged and neutralized the enemy effectively.

The conflict witnessed several firsts for air defense, including the strategic maneuver of Patriot forces to the front lines and the largest movement of Patriot forces by air or land in history. Those achievements, along with the innovative tactics deployed, would later serve as pivotal reasons for the establishment of future departments dedicated to commemorating those accomplishments. Those new departments would also address and devise solutions for the challenges encountered by the Patriot force in the harsh desert conditions of Iraq, ensuring that lessons learned would enhance future operations and strategies.

While still in Baghdad, continuing our training based on the existing military doctrines, we finally received the orders we had been eagerly anticipating—to move out and head back to Kuwait. There, we would await further instructions to return home. The mood was jubilant; the news was celebrated in a manner that transformed our routine. Meals Ready to Eat, typically consumed straight from their standard brown packaging, were instead enjoyed on Styrofoam plates, adding a touch of normalcy to our dining experience.

The celebration extended into the evening with a movie night, a rare treat in our environment. Many of us gathered around a small computer screen, enjoying episodes of CSI from DVDs that had been sent in a care package. That simple pleasure brought us a slice of home and camaraderie, reinforcing the reality that we were indeed going home—a moment we all cherished deeply.

الرحلات القادمة

بغداد
Baghdad
بناية المسافرين أ
ابنية الخدمات
Passenger Terminal A
Service Buildings

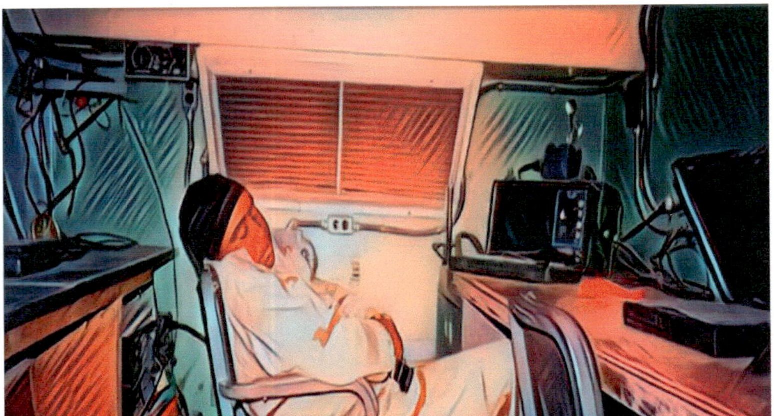

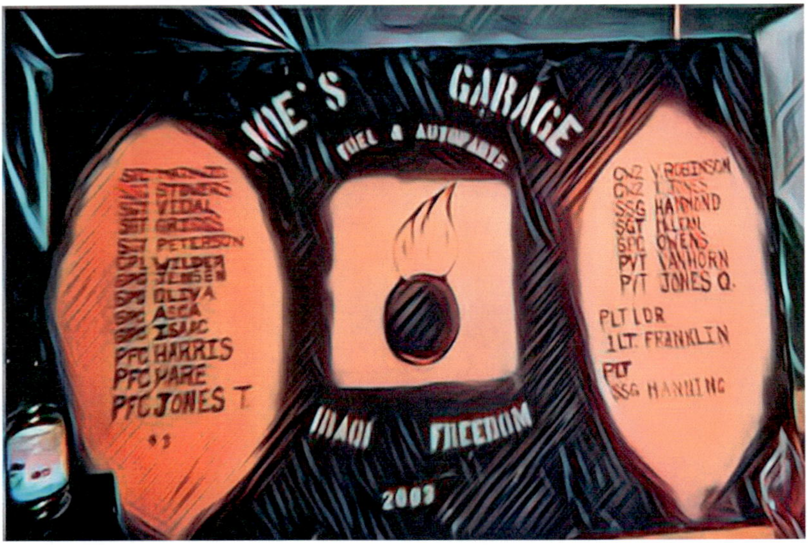

JOE'S GARAGE
FUEL & AUTOPARTS
VIDAL
PETERSON
WILDER
JENSEN
OLIVA
ISAAC
PFC HARRIS
PFC JONES T.
SSG HAMMOND
SPC OWENS
PVT VANHORN
JONES Q.
PLT LDR
1LT. FRANKLIN
SSG MANNING
IRAQI FREEDOM

Movement Back to Kuwait

Camp Begin Bound.

In the early days of June 2003, we made our way back to Camp Begin in Kuwait.

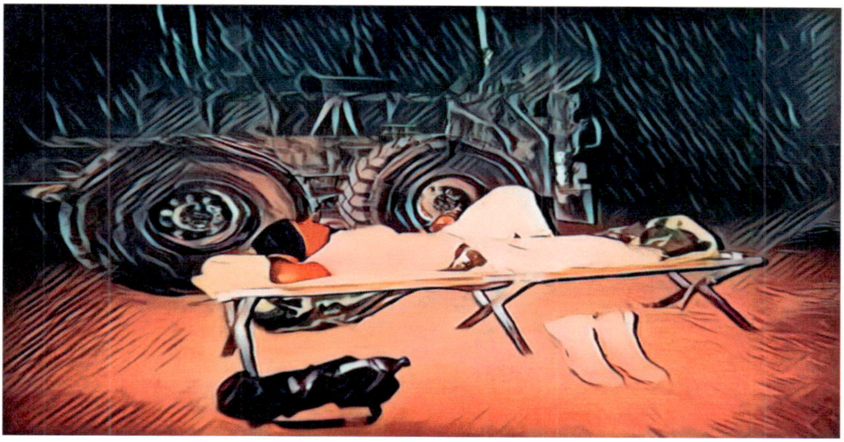

On the day we set out for Camp Begin, our journey spanned approximately nine hours, punctuated only by necessary stops to repair broken equipment. During our transit, our convoy was attacked by a group of Iraqi civilians, resulting in a minor injury to one team member. Despite that incident, we were able to continue on our path toward Kuwait. That evening, we reached another camp, characterized mainly by its dusty environment, where we settled for the night.

The following morning, we resumed our journey at 6:30 am and successfully arrived at Camp Begin around 4:00 pm, coinciding with the arrival of the last vehicle in our convoy. Upon arrival, we conducted a thorough check of our vehicles and settled down for the night. The roads during our return were notably calm, showing little activity, which marked a stark contrast to the earlier part of our travels. The calmness allowed us to focus on the road ahead and the anticipation of eventually heading home.

After successfully completing our redeployment to Camp Begin, our immediate plan was to continue training while maintaining alert state zero, as we awaited further orders to redeploy to our home station in Germany. The termination of major combat operations and the assessed neutralization of the Iraqi air threat led the Central Command to officially release our air defense forces from their duties covering the area skies. That pivotal decision marked a significant transition for our unit, signaling a return to more routine duties and preparation for eventual departure.

Soon after, we received the anticipated redeployment orders to Germany. Following those orders, we moved our equipment to the port. There, it was methodically prepared and loaded for shipment back to Germany, setting the stage for our return to our home base and concluding our active deployment in the region. The process not only signified the end of our operational role in Iraq but also the beginning of reintegration and reflection on our contributions during a critical period of military engagement.

After transporting all of our weapon system equipment to the port, we continued to stay at Camp Begin for many days, enduring the sweltering Kuwaiti heat. Despite the challenging conditions, our spirits remained high. We had previously lived in the most austere conditions in the battle zones of Iraq, before any established camps were built, lacking even makeshift showers or latrines. We had consistently operated ahead of basic comforts.

Now, being back in Kuwait, equipped with amenities such as latrines, showers, a tent over our heads, and access to hot food felt like a significant upgrade. That was a luxury compared to the harsh environments we had endured. Our team had not only met every demand asked of us but had gone beyond, accomplishing feats never before achieved by any Patriot unit, all under the most challenging circumstances with limited resources.

Our ability to adapt and make do with the resources available on the battlefield, including utilizing oversized vehicles for various needs, was remarkable. The ingenuity made coping with the intense heat of the Kuwaiti desert more manageable. As a result, despite the discomfort, our resilience and adaptability made the situation bearable, even in such extreme conditions.

By the end of June 2003, all Patriot air defense forces, with the exception of those involved in routine rotations in the region, had been redeployed to their home stations. That marked a significant milestone in the drawdown of specialized military assets following the major combat operations in Iraq. However, despite the withdrawal of those specific forces, stabilization operations continued across Iraq.

The Return Home

The Return from Battle.

Homecoming was a profound relief, yet it was not without its burdens. The blessing was clear: everyone who had deployed with us returned home. Nonetheless, we carried back more than just physical baggage; the experiences had left a lasting imprint on us.

The conflict had begun in earnest on March 20, 2003, when Iraqi forces launched missiles at coalition troops stationed in Kuwait. Those missiles were quickly detected, triggering immediate alarms that warned of the looming danger. That moment marked the beginning of intense engagement for American troops, thrusting them into the heart of conflict—a pivotal event that would change each of us in various ways.

Returning home from conflict was a profound and complex experience, often accompanied by feelings that weren't immediately recognizable or easy to articulate. The sadness and depression described was common among veterans who had faced intense situations and stressors that were far removed from civilian life. But recognizing those feelings was particularly challenging, especially in a military culture that historically emphasized resilience and toughness over vulnerability.

The reintegration sessions we attended were meant to aid in the transition, but the prevailing attitude of "suck it up" rather than “seek help” made it difficult to address those emotional and psychological challenges openly. It wasn’t uncommon for service members to deny feelings of depression or sadness, either because they were unaware of them or because they felt pressured to appear unaffected by their experiences.

I would later learn that feeling different and not quite knowing how to fit back into our previous reality was an entirely valid response to what we’d been through. It was a testament to the profound impact of our experiences overseas. Acknowledging that something had changed within us was a crucial first step toward understanding our feelings and finding a way to heal.

The transition from the constant hyper-vigilance required in a conflict zone like Iraq to the relative peace of home was jarring and at times difficult. In Iraq, we were constantly alert, always ready to respond to immediate threats and maintain high operational standards.

That state of heightened awareness was crucial in a war zone but felt out of place back home, where the environment no longer required such intense vigilance; I was changed.

The displacement that I was feeling would remain for some time but would eventually subside and I would regain some joy and happiness and the desire to travel again; all of which was taken from me after returning from the conflict. I would spend another couple of years at the duty station in Germany before departing for my next duty assignment.

The early years of combat operations in Iraq were marked by intense experiences and powerful stories, many of which were harrowing and highlighted the grim realities of war. The impact of those operations stretched broadly across the U.S. military, touching the lives of nearly every service member. Over 300,000 men and women in the Armed Services participated in that operation, each carrying home their own personal stories from their time in a war-torn land. The collective experience formed a significant chapter in America's ongoing war on terrorism.

The events of September 11, 2001, had a profound effect on the national psyche, triggering a wide range of emotions from anger and fear to a determined call for action. That tragic day shifted the way Americans viewed their safety and security, leading to heightened vigilance and a pervasive sense of vulnerability. The need to retaliate and protect the nation not only led to military responses abroad but also transformed domestic attitudes, fostering a climate where suspicion and fear of potential threats became more commonplace.

That period in history was a transformative time for the United States, reshaping policies, military practices, and the everyday lives of countless individuals. The changes it brought about continue to influence discussions on national security, foreign policy, and how the nation responds to international threats.

The narrative of sacrifice for freedom was deeply etched in the experiences of soldiers, sailors, marines, and airmen who confronted the harsh realities of war firsthand. Witnessing the devastation of communities and the visible despair on the faces of civilians amid the ruins, profoundly affected those serving in the conflict zones. The impact of such scenes extended far beyond the battlefield, touching the very core of what it meant to fight for and defend a way of life.

As the conflict wore on, the soldiers were exposed to increasingly horrific scenes. They encountered the grim aftermath of battles—bodies in buildings and wounded comrades—images that were not easily forgotten. Those experiences were compounded by personal loss, as many service members carried the weight of stories about fellow comrades who had fallen in combat. Those stories often included visceral details of violence and loss, painting a stark picture of the realities of war.

Moreover, those service members faced constant risks, from the fear of captivity to the relentless uncertainty of not returning home. The emotional and psychological burden of advancing into battle in a land that offered no promises, added another layer of hardship.

Such experiences highlight the immense personal sacrifices made by those who serve. The stories of those men and women were more than just accounts of survival; they were poignant reminders of the human cost of war and the profound resilience required to navigate such extreme circumstances.

The journey of my air defense unit, along with many others, would uncover stories similar to ours as we deployed to the region just before combat operations commenced. We joined in support of the mission. Our tale might not be the stuff of movies or the subject of fervent newspaper articles, but it was deemed "remarkable" by our Command, earning us praise.

My new unit and I were called to serve our country more than a year after the initial attacks on 9/11. Although our contribution might have seemed minor, we formed an organization comprised of many individuals who had never faced conflict or war, and many were new to the military, never anticipating what fighting really entailed. We were a team of soldiers—initially frightened yet simultaneously excited and eager—who wanted to alleviate the nation's suffering. We aimed to defend our way of life and respond to the call of patriotism. That was why we engaged in numerous field exercises, training events, and skill certifications, ensuring we were prepared should we be called into action.

Before the conflict began, there were no guarantees that we would encounter direct action, nor were there assurances that we would emerge unscathed or even survive. Furthermore, there was no certainty that we wouldn't be profoundly changed by our experiences—a reality many would come to face. When the order was issued, the potential impact on our lives was unknown, but we were prepared to accept the mission.

Initially, my time in the military was intended to be just a two-year enlistment, primarily because I had exceeded the age limit to enter as an officer and thus had to enlist as a soldier. My decision to join as an enlisted member was driven largely by a sense of compulsion; I had met all the entry requirements, and at the time, I harbored no expectations of participating in actual combat.

What began as an intended two-year stint in the military unfolded into a fulfilling twenty-year career, a path I hadn't originally planned on taking. When I first enlisted, despite learning that I couldn't become an officer, it was a compelling call to serve, and that call was irresistible.

Year after year, I enjoyed many duty assignments. I was grateful for the individuals I served with and those I met along the way. I basked in the good fortune of being able to travel to locations that I would never have gotten to without being in the military, and I celebrated the growth I experienced as a person. I had absolutely no regrets about my 20-year stint, not even Iraq.

As the sun sets in the desert lands of Iraq!

Made in the USA
Las Vegas, NV
07 November 2026

57669925R00050